Name: Eng. /
Mostafa Yacoub Abdellatif Mahmoud
Nationality: Egyptian
ORCID: 0000-0002-9991-4624
Email:
moshhaabma2015@gmail.com
Qualification: civil engineer Cairo
University 2003

- regularity within the distribution of prime numbers

  In this paper or research, we will explain the regularity within the distribution of prime numbers based on my discovered formula that connects prime and composite numbers.

My discovered formula:

<u>Definitions:</u>

Array PTBP

It is the following Array of odd numbers

$$
\begin{vmatrix}
1 & 3 & 7 & 9 \\
11 & 13 & 17 & 19 \\
21 & 23 & 27 & 29 \\
31 & 33 & 37 & 39 \\
41 & 43 & 47 & 49 \\
51 & 53 & 57 & 59
\end{vmatrix}
$$

And so on....

- For a given set of consecutive primes whose numbers =n that start with prime 3 and end with prime F and not including prime 2 and prime 5 i.e.

set=[3,7,11,13,…………………………………
………………………..,F]

S=product of those consecutive primes

i.e

$$S = \prod_{i=3}^{i=F} (i)$$

Range=R_k = 10 × S×k

Where k = [1, 2, 3, 4, ...,
∞(infinity)

i.e R_1=10 x S x 1 and R_2=10 x S x 2

And so on

- Number of composite numbers that belong to Array PTBP and created by the effect of those consecutive primes within the range R_K

- $=[(K \times 4^{\times \frac{S}{3}}) + ($

$$\sum_{j=7}^{j=F} (K \times 4 \times (\frac{S}{j}) \times$$

$i = $ *prime number befor current prime number j*

$$\prod_{i=7} \qquad (\frac{i-1}{i})$$

)]-(n)

Where j =consecutive values of primes

7, 11, 13,............., F

And i= consecutive values of primes

3, 7, 11, 13,........, prime before current j prime

The previous formula can be applied for any number of consecutive prime numbers that start with prime number 3

- The first term $(k \times 4 \times \frac{S}{3})$ represents the count of unique Composite numbers +1 that belong to the Array PTBP and are created by prime number 3 within the range

$$R_k = 10 \times S \times k$$

- **The second term**

$$\sum_{j=7}^{j=F} (K \times 4 \times (\frac{S}{j}) \times$$

$i =$ prime number befor current prime number j

$$\prod_{i=7} (\frac{i-1}{i})$$

Represent the count of unique Composite numbers+n-1 that belong to the Array PTBP and are created by each prime number after the prime number 3 within the range

$R_k = 10 \times S \times k$

- **The third term (-n)**
Subtracting n (number of consecutive primes starting from prime number 3) because the count of composite numbers generated from those consecutive primes includes the count of those primes in the range

$R_k = 10 \times S \times k$

- **Explanation and proof for my theory in my previous paper (prime number theory)**
- **We will mention only the concept of number cycle**

  We can use the number cycle concept to understand the behavior of consecutive primes in creating composite numbers.

  i.e.

  $$S = \prod_{i=3}^{i=F} (i)$$

  Range=cycle range= R_k = 10 × S × k

  Where k= [1, 2, 3, 4, ..., ∞(infinity)

  i.e. R_1=10 x S x 1 and R_2=10 x S x 2

  And so on

- **Now consider only one k value =1**

For a set of consecutive primes and according to my formula the result will be

$$=[(K \times 4^{\times \frac{S}{3}}) + ($$

$$\sum_{j=7}^{j=F} (K \times 4 \times (\frac{S}{j}) \times$$

$i = prime\ number\ before\ current\ prime\ number\ j$

$$\prod_{i=7} (\frac{i-1}{i})$$

$$)]-(n)$$

And

including the count of prime numbers within the set

$$=[(K \times 4^{\times \frac{S}{3}}) + ($$

$$\sum_{j=7}^{j=F} (K \times 4 \times (\frac{S}{j}) \times$$

$i = prime\ number\ before\ current\ prime\ number\ j$

$$\prod_{i=7} (\frac{i-1}{i})$$

$$)]$$

- Which represent the count of numbers (that belong to the Array PTBP) that is divisible of the prime numbers that belong to the set of consecutive primes

- ^{t}he complementary part

$$= (4 \times \prod_{i=3}^{i=F} (i-1))$$

  as explained in my paper (infinite primes)

- Which represent the count of numbers (that belong to the Array PTBP) that is not divisible by each prime number that belong to the set of consecutive primes

- For example if the set of consecutive primes=[3] then we have the following pattern of uncolored numbers (except 1) that continues to infinity

- and we can Define and limit the scope of the search for prime numbers within that repeated pattern

1	3	7	9
11	13	17	19
21	23	27	29
31	33	27	39
41	43	27	49
51	53	27	59
61	63	27	69
71	73	27	79
81	83	27	89
91	93	27	99
101	103	27	109
111	113	27	119

- And this pattern continues to infinity
- And now if the set of consecutive primes=[3,7] then we have the following more accurate repeated pattern of uncolored numbers (except 1) that continues to infinity
- Which limits the process of searching for prime numbers to more precise places
- than the previous set that has only the prime number 3

1	3	7	9
11	13	17	19
21	23	27	29

31	33	37	39
41	43	47	**49**
51	53	57	59
61	63	67	69
71	73	**77**	79
81	83	87	89
91	93	97	99
101	103	107	109
111	113	117	**119**
121	123	127	129
131	**133**	137	139
141	143	147	149
151	153	157	159
161	163	167	169
171	173	177	179

181	183	187	189
191	193	197	199
201	203	207	209
211	213	217	219
221	223	227	229
231	233	237	239
241	243	247	249
251	253	257	259
261	263	267	269
271	273	277	279
281	283	287	289
291	293	297	299
301	303	307	309
311	313	317	319
321	323	327	329

331	333	337	339
341	343	347	349
351	353	357	359
361	363	367	369
371	373	377	379
381	383	387	389
391	393	397	399
401	403	407	409
411	413	417	419
421	423	427	429
431	433	437	439
441	443	447	449
451	453	457	459
461	463	467	469
471	473	477	479

481	483	487	489
491	493	497	499
501	503	507	509
511	513	517	519
521	523	527	529
531	533	537	539
541	543	547	549
551	553	557	559
561	563	567	569
571	573	577	579
581	583	587	589
591	593	597	599
601	603	607	609
611	613	617	619
621	623	627	629

- And this pattern continues to infinity
- As we add more consecutive prime numbers to the set we get a more accurate pattern of uncolored numbers that continues to infinity Which limits the process of searching for prime numbers to more precise places

- and the percentage of those uncolored numbers decreases relative to the range of that set for all values of k from 1 to infinity

- And we can define that percentage to be equal to

$$[(4 \times \prod_{i=3}^{i=F} (i-1))] / [4 \times \prod_{i=3}^{i=F} (i)]$$

$$= [(\prod_{i=3}^{i=F} (i-1))] / [\prod_{i=3}^{i=F} (i)]$$

$$= \prod_{i=3}^{i=F} [(i-1)/i]$$

Where F is the last prime number in the set

- And now if we take the set of consecutive primes=[3,7,11]

The first cycle will be as follows table

With range = 3 x 7 x 11 x 10=2310

Number of unique composite numbers produced by prime number 11

=4 x 2 x6 =48

Or

= 4 x [3x7x11 / 11] x ((2/3) x (6/7)) = 48 including prime 11 itself within the first cycle (k=1) and 48 composite numbers in each following cycle i.e for K>1

1	3	7	9
11	13	17	19
21	23	27	29
31	33	37	39
41	43	47	49
51	53	57	59
61	63	67	69
71	73	77	79
81	83	87	89
91	93	97	99
101	103	107	109
111	113	117	119
121	123	127	129
131	133	137	139
141	143	147	149
151	153	157	159
161	163	167	169
171	173	177	179
181	183	187	189
191	193	197	199
201	203	207	209

211	213	217	219
221	223	227	229
231	233	237	239
241	243	247	249
251	253	257	259
261	263	267	269
271	273	277	279
281	283	287	289
291	293	297	299
301	303	307	309
311	313	317	319
321	323	327	329
331	333	337	339
341	343	347	349
351	353	357	359
361	363	367	369
371	373	377	379
381	383	387	389
391	393	397	399
401	403	407	409
411	413	417	419

421	423	427	429
431	433	437	439
441	443	447	449
451	453	457	459
461	463	467	469
471	473	477	479
481	483	487	489
491	493	497	499
501	503	507	509
511	513	517	519
521	523	527	529
531	533	537	539
541	543	547	549
551	553	557	559
561	563	567	569
571	573	577	579
581	583	587	589
591	593	597	599
601	603	607	609
611	613	617	619
621	623	627	629

631	633	637	639
641	643	647	649
651	653	657	659
661	663	667	669
671	673	677	679
681	683	687	689
691	693	697	699
701	703	707	709
711	713	717	719
721	723	727	729
731	733	737	739
741	743	747	749
751	753	757	759
761	763	767	769
771	773	777	779
781	783	787	789
791	793	797	799
801	803	807	809
811	813	817	819
821	823	827	829
831	833	837	839

841	843	847	849
851	853	857	859
861	863	867	869
871	873	877	879
881	883	887	889
891	893	897	899
901	903	907	909
911	913	917	919
921	923	927	929
931	933	937	939
941	943	947	949
951	953	957	959
961	963	967	969
971	973	977	979
981	983	987	989
991	993	997	999
1001	1003	1007	1009
1011	1013	1017	1019
1021	1023	1027	1029
1031	1033	1037	1039
1041	1043	1047	1049

1051	1053	1057	1059
1061	1063	1067	1069
1071	1073	1077	1079
1081	1083	1087	1089
1091	1093	1097	1099
1101	1103	1107	1109
1111	1113	1117	1119
1121	1123	1127	1129
1131	1133	1137	1139
1141	1143	1147	1149
1151	1153	1157	1159
1161	1163	1167	1169
1171	1173	1177	1179
1181	1183	1187	1189
1191	1193	1197	1199
1201	1203	1207	1209
1211	1213	1217	1219
1221	1223	1227	1229
1231	1233	1237	1239
1241	1243	1247	1249
1251	1253	1257	1259

1261	1263	1267	1269
1271	1273	1277	1279
1281	1283	1287	1289
1291	1293	1297	1299
1301	1303	1307	1309
1311	1313	1317	1319
1321	1323	1327	1329
1331	1333	1337	1339
1341	1343	1347	1349
1351	1353	1357	1359
1361	1363	1367	1369
1371	1373	1377	1379
1381	1383	1387	1389
1391	1393	1397	1399
1401	1403	1407	1409
1411	1413	1417	1419
1421	1423	1427	1429
1431	1433	1437	1439
1441	1443	1447	1449
1451	1453	1457	1459
1461	1463	1467	1469

1471	1473	1477	1479
1481	1483	1487	1489
1491	1493	1497	1499
1501	1503	1507	1509
1511	1513	1517	1519
1521	1523	1527	1529
1531	1533	1537	1539
1541	1543	1547	1549
1551	1553	1557	1559
1561	1563	1567	1569
1571	1573	1577	1579
1581	1583	1587	1589
1591	1593	1597	1599
1601	1603	1607	1609
1611	1613	1617	1619
1621	1623	1627	1629
1631	1633	1637	1639
1641	1643	1647	1649
1651	1653	1657	1659
1661	1663	1667	1669
1671	1673	1677	1679

1681	1683	1687	1689
1691	1693	1697	1699
1701	1703	1707	1709
1711	1713	1717	1719
1721	1723	1727	1729
1731	1733	1737	1739
1741	1743	1747	1749
1751	1753	1757	1759
1761	1763	1767	1769
1771	1773	1777	1779
1781	1783	1787	1789
1791	1793	1797	1799
1801	1803	1807	1809
1811	1813	1817	1819
1821	1823	1827	1829
1831	1833	1837	1839
1841	1843	1847	1849
1851	1853	1857	1859
1861	1863	1867	1869
1871	1873	1877	1879
1881	1883	1887	1889

1891	1893	1897	1899
1901	1903	1907	1909
1911	1913	1917	1919
1921	1923	1927	1929
1931	1933	1937	1939
1941	1943	1947	1949
1951	1953	1957	1959
1961	1963	1967	1969
1971	1973	1977	1979
1981	1983	1987	1989
1991	1993	1997	1999
2001	2003	2007	2009
2011	2013	2017	2019
2021	2023	2027	2029
2031	2033	2037	2039
2041	2043	2047	2049
2051	2053	2057	2059
2061	2063	2067	2069
2071	2073	2077	2079
2081	2083	2087	2089
2091	2093	2097	2099

2101	2103	2107	2109
2111	2113	2117	2119
2121	2123	2127	2129
2131	2133	2137	2139
2141	2143	2147	2149
2151	2153	2157	2159
2161	2163	2167	2169
2171	2173	2177	2179
2181	2183	2187	2189
2191	2193	2197	2199
2201	2203	2207	2209
2211	2213	2217	2219
2221	2223	2227	2229
2231	2233	2237	2239
2241	2243	2247	2249
2251	2253	2257	2259
2261	2263	2267	2269
2271	2273	2277	2279
2281	2283	2287	2289
2291	2293	2297	2299
2301	2303	2307	2309

- Numbers with yellow color represent numbers that are divisible by prime number 3.
- 21 counts for prime number 3 not for prime number 7 because prime number 3 has a priority.
- Numbers with green color represent numbers that are divisible by prime number 7.
- Numbers with blue color represent numbers that are divisible by prime number 11.
- Count of uncolored numbers
  $= 4 \times 2 \times 6 \times 10 = 480$
- Each column of the array PTBP represents the produced cycle by the set of consecutive primes [3,7,11]
  Has a count $= 2 \times 6 \times 10 = 120$

- **The sum of uncolored numbers in each column which summarized in the following table=138600**

1	13	17	19
31	23	37	29
41	43	47	59
61	53	67	79
71	73	97	89
101	83	107	109
131	103	127	139
151	113	137	149
181	163	157	169
191	173	167	179
211	193	197	199
221	223	227	229
241	233	247	239
251	263	257	269
271	283	277	289
281	293	307	299
311	313	317	349
331	323	337	359
361	353	347	379
391	373	367	389
401	383	377	409
421	403	397	419
431	433	437	439
461	443	457	449
481	463	467	479
491	493	487	499
521	503	527	509
541	523	547	529
551	533	557	559
571	563	577	569
601	593	587	589
611	613	607	599
631	643	617	619
641	653	647	629
661	673	667	659
691	683	677	689
701	703	697	709
731	713	727	719
751	733	757	739
761	743	767	769

811	773	787	779
821	793	797	799
841	823	817	809
851	853	827	829
871	863	857	839
881	883	877	859
901	893	887	899
911	923	907	919
941	943	937	929
961	953	947	949
971	983	967	989
991	1003	977	1009
1021	1013	997	1019
1031	1033	1007	1039
1051	1063	1027	1049
1061	1073	1037	1069
1081	1093	1087	1079
1091	1103	1097	1109
1121	1123	1117	1129
1151	1153	1147	1139
1171	1163	1157	1159
1181	1193	1187	1189
1201	1213	1207	1219
1231	1223	1217	1229
1241	1273	1237	1249
1261	1283	1247	1259
1271	1303	1277	1279
1291	1313	1297	1289
1301	1333	1307	1319
1321	1343	1327	1339
1361	1363	1357	1349
1381	1373	1367	1369
1391	1403	1387	1399
1411	1423	1417	1409
1451	1433	1427	1429
1471	1453	1447	1439
1481	1483	1457	1459
1501	1493	1487	1469
1511	1513	1517	1489
1531	1523	1537	1499

1541	1543	1567	1549
1571	1553	1577	1559
1591	1583	1597	1579
1601	1613	1607	1609
1621	1633	1627	1619
1651	1643	1637	1649
1681	1663	1657	1669
1691	1693	1667	1679
1711	1703	1697	1699
1721	1723	1717	1709
1741	1733	1747	1739
1751	1753	1777	1759
1781	1763	1787	1769
1801	1783	1807	1789
1811	1823	1817	1819
1831	1843	1847	1829
1861	1853	1867	1849
1871	1873	1877	1879
1891	1913	1907	1889
1901	1933	1927	1909
1921	1943	1937	1919
1931	1963	1957	1949
1951	1973	1987	1979
1961	1993	1997	1999
2011	2003	2017	2029
2021	2033	2027	2039
2041	2053	2047	2059
2071	2063	2077	2069
2081	2083	2087	2089
2111	2113	2117	2099
2131	2143	2137	2119
2141	2153	2147	2129
2161	2173	2197	2159
2171	2183	2207	2179
2201	2203	2227	2209
2221	2213	2237	2239
2231	2243	2257	2249
2251	2263	2267	2269
2281	2273	2287	2279
2291	2293	2297	2309
138600	138600	138600	138600

- We have proved that the sum of all uncolored numbers must(prove exist in my book (prime number theory part 2))=

$$=20 \times \prod_{i=3}^{i=11} [i \times (i-1)]$$

- Here we have uniform distribution among the four columns of the array ptbp produced by the first cycle of the set of consecutive primes [3,7,11]

So the sum of each column must be

$$=5 \times \prod_{i=3}^{i=11} [i \times (i-1)]$$

$$=5 \times 2 \times 3 \times 6 \times 7 \times 10 \times 11 = 138600$$

- The next prime number must produce the equal sum of the produced composite numbers within the first cycle(including the value of the prime number itself within the first cycle) Of the Set=[3,7,11, P]

- We have proved that the sum of composite numbers produced by prime P within the first cycle including the prime p itself self must

$$= 20 \times P \times \prod_{i=3}^{i=11} [i \times (i-1)]$$

It must be uniformly distributed among the four columns of the array ptbp produced by the set of consecutive primes [3,7,11, P]

- So it must be $= 5 \times P \times \prod_{i=3}^{i=11} [i \times (i-1)]$

$=5 \times P \times [2 \times 3 \times 6 \times 7 \times 10 \times 11]$ each column

And the sum of uncolored numbers (numbers that are not divisible by each of prime numbers 3,7,11, P) in each column within the first cycle of the set[3,7,11, P]

Must=

$$(138600 \times P) + [(5 \times P \times (P-1)$$

$$\times \prod_{i=3}^{i=11} [i \times (i-1)] \,] - [\, 5 \times P$$

$$\times \prod_{i=3}^{i=11} [i \times (i-1)] \,]$$

- And any next cycle produced by any next set of consecutive primes must have the uniform distribution of the summation of uncolored numbers(numbers that are not divisible by each of the consecutive primes that belong to the set) among each column of the array ptbp

- This uniform distribution of the summation of uncolored numbers continues for other values of $K > 1$

- Because each uncolored number (let value =A) in any column in the first cycle has an instance in each of the next cycle
  with value

$$=A +[(K-1)^{x\,10\,x\,\prod_{i=3}^{i=F}[i]}]$$

- F = the value of the last prime number within the set of consecutive primes
  Where k= [1, 2, 3, 4, ..., ∞(infinity)

www.ingramcontent.com/pod-product-compliance
Lightning Source LLC
Chambersburg PA
CBHW071018290526
45795CB00005B/1854